CHAPTER 3

CIRCUIT CONTROL DEVICES

LEARNING OBJECTIVES

Upon completion of this chapter you will be able to:

1. State three reasons circuit control devices are used and list three general types of circuit control devices.

2. Identify the schematic symbols for a switch, a solenoid, and a relay.

3. State the difference between a manual and an automatic switch and give an example of each.

4. State the reason multicontact switches are used.

5. Identify the schematic symbols for the following switches:

    *   Single-pole, double-throw

    *   Double-pole, single-throw

    *   Double-pole, double-throw

    *   Single-break

    *   Double-break

    *   Rotary

    *   Wafer

6. State the characteristics of a switch described as a rocker switch.

7. State the possible number of positions for a single-pole, double-throw switch.

8. Identify a type of momentary switch.

9. State the type of switch used to prevent the accidental energizing or deenergizing of a circuit.

10. State the common name for an accurate snap-acting switch.

11. State the meaning of the current and voltage rating of a switch.

12. State the two types of meters you can use to check a switch.

13. Select the proper substitute switch from a list.

14. State the conditions checked for in preventive maintenance of switches.

15. State the operating principle and one example of a solenoid.

16. State the ways in which a solenoid can be checked for proper operation.

17. State the operating principle of a relay and how it differs from a solenoid.

18. State the two types of relays according to use.

19. State the ways in which a relay can be checked for proper operation and the procedure for servicing it.

CIRCUIT CONTROL DEVICES

Circuit control devices are used everywhere that electrical or electronic circuits are used. They are found in submarines, computers, aircraft, televisions, ships, space vehicles, medical instruments, and many other places. In this chapter you will learn what circuit control devices are, how they are used, and some of their characteristics.

INTRODUCTION

Electricity existed well before the beginning of recorded history. Lightning was a known and feared force to early man, but the practical uses of electricity were not recognized until the late 18th century. The early experimenters in electricity controlled power to their experiments by disconnecting a wire from a battery or by the use of a clutch between a generator and a steam engine. As practical uses were found for electricity, a convenient means for turning power on and off was needed.

Telegraph systems, tried as early as the late 1700s and perfected by Morse in the 1830s, used a mechanically operated contact lever for opening and closing the signal circuit. This was later replaced by the hand-operated contact lever or "key."

Early power switches were simple hinged beams, arranged to close or open a circuit. The blade-and-jaw knife switch with a wooden, slate, or porcelain base and an insulated handle, was developed a short time later. This was the beginning of circuit control devices.

Modern circuit control devices can change their resistance from a few milliohms (when closed) to well over 100,000 megaohms (when open) in a couple of milliseconds. In some circuit control devices, the movement necessary to cause the device to open or close is only .001 inch (.025 millimeters).

NEED FOR CIRCUIT CONTROL

Circuit control, in its simplest form, is the application and removal of power. This can also be expressed as turning a circuit on and off or opening and closing a circuit. Before you learn about the types of circuit control devices, you should know why circuit control is needed.

If a circuit develops problems that could damage the equipment or endanger personnel, it should be possible to remove the power from that circuit. The circuit protection devices discussed in the last chapter will remove power automatically if current or temperature increase enough to cause the circuit protection device to act. Even with this protection, a manual means of control is needed to allow you to remove power from the circuit before the protection device acts.

When you work on a circuit, you often need to remove power from it to connect test equipment or to remove and replace components. When you remove power from a circuit so that you can work on it, be sure to "tag out" the switch to ensure that power is not applied to the circuit while you are working. When work has been completed, power must be restored to the circuit. This will allow you to check the proper operation of the circuit and place it back in service. After the circuit has been checked for proper operation, remove the tag from the power switch.

Many electrical devices are used some of the time and not needed at other times. Circuit control devices allow you to turn the device on when it is needed and off when it is not needed.

Some devices, like multimeters or televisions, require the selection of a specific function or circuit. A circuit control device makes possible the selection of the particular circuit you wish to use.

TYPES OF CIRCUIT CONTROL DEVICES

Circuit control devices have many different shapes and sizes, but most circuit control devices are either SWITCHES, SOLENOIDS, or RELAYS.

Figure 3-1 shows an example of each of these types of circuit control devices and their schematic symbols.

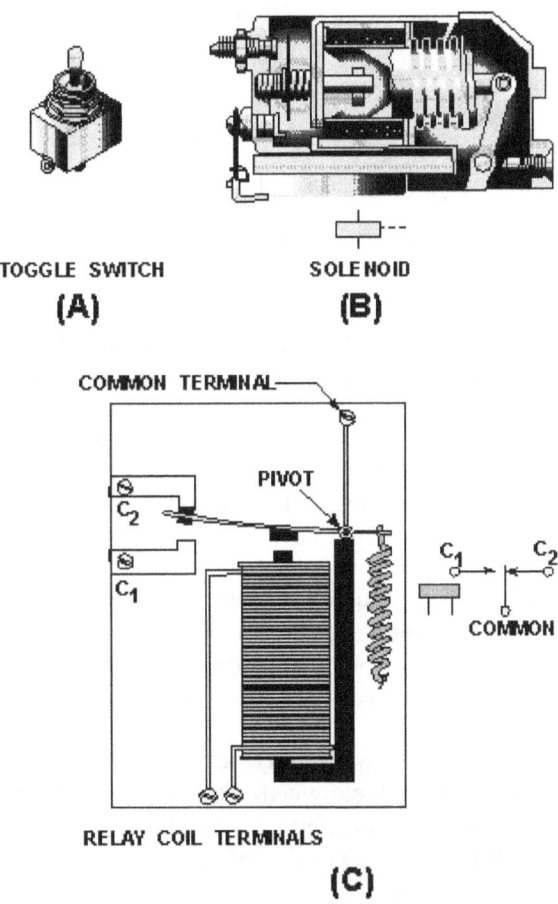

Figure 3-1.—Typical circuit control devices: RELAY COIL TERMINALS

Figure 3-1, view A, is a simple toggle switch and the schematic symbol for this switch is shown below it. Figure 3-1, view B, is a cutaway view of a solenoid. The schematic symbol below the solenoid is one of the schematic symbols used for this solenoid. Figure 3-1, view C, shows a simple relay. One of the schematic symbols for this relay is shown next to the relay.

Q1. What are three reasons circuit control is needed?

Q2. What are the three types of circuit control devices?

Q3. Label the schematic symbols shown in figure 3-2.

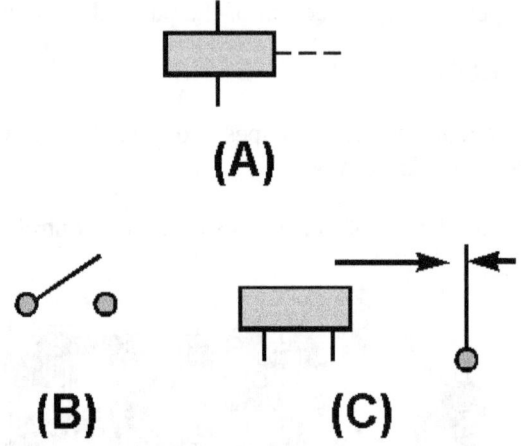

Figure 3-2.—Schematic symbol recognition.

SWITCH TYPES

There are thousands and thousands of switch applications found in home, industry, and the Navy. Hundreds of electrical switches work for you everyday to perform functions you take for granted. Some switches operate by the touch of a finger and many others are operated automatically.

Switches are used in the home to turn off the alarm clock, to control the stove, to turn on the refrigerator light, to turn on and control radios and televisions, hair dryers, dishwashers, garbage disposals, washers and dryers, as well as to control heating and air conditioning. A typical luxury automobile with power seats and windows might have as many as 45 switches.

Industry uses switches in a wide variety of ways. They are found in the business office on computers, copy machines, electric typewriters, and other equipment. A factory or shop may use thousands of switches and they are found on almost every piece of machinery. Switches are used on woodworking machinery, metal working machinery, conveyors, automation devices, elevators, hoists, and lift trucks.

The Navy uses switches in a number of ways. A typical aircraft could have over 250 switches to control lights, electronic systems, and to indicate whether the landing gear is up or down. Ships, fire control systems, and missile launchers are also controlled by electrical switches. In fact, almost all electrical or electronic devices will have at least one switch.

Switches are designed to work in many different environments from extreme high pressure, as in a submarine, to extreme low pressure, as in a spacecraft. Other environmental conditions to consider are high or low temperature, rapid temperature changes, humidity, liquid splashing or immersion, ice, corrosion, sand or dust, fungus, shock or vibration, and an explosive atmosphere.

It would not be possible to describe all the different switches used. This chapter will describe the most common types of switches.

MANUAL SWITCHES

A manual switch is a switch that is controlled by a person. In other words, a manual switch is a switch that you turn on or off. Examples of common manual switches are a light switch, the ignition switch on a motor vehicle, or the channel selector on a television. You may not think of the channel selector as a switch that you use to turn something on or off, but that is what it does. The channel selector is used to turn on the proper circuit and allows the television to receive the channel you have selected.

AUTOMATIC SWITCH

An automatic switch is a switch that is controlled by a mechanical or electrical device. You do not have to turn an automatic switch on or off. Two examples of automatic switches are a thermostat and the distributor in a motor vehicle. The thermostat will turn a furnace or air conditioner on or off by responding to the temperature in a room. The distributor electrically turns on the spark plug circuit at the proper time by responding to the mechanical rotation of a shaft. Even the switch that turns on the light in a refrigerator when the door is opened is an automatic switch.

Automatic switches are not always as simple as the examples given above. Limit switches, which sense some limit such as fluid level, mechanical movement, pressure (altitude or depth under water), or an electrical quantity, are automatic switches. Computers use and control automatic switches that are sometimes quite complicated.

Basically, any switch that will turn a circuit on or off without human action is an automatic switch.

MULTICONTACT SWITCHES

Switches are sometimes used to control more than one circuit or to select one of several possible circuits. An example of a switch controlling more than one circuit is the AM/FM selector on a radio. This switch enables you to control either the AM or FM portion of the radio with a single switch. An example of a switch that selects one of several circuits is the channel selector of a television set. These switches are called MULTICONTACT switches because they have more than one contact or MULTI(ple) CONTACTS.

Number of Poles and Number of Throws

Multicontact switches (other than rotary switches, which will be covered later) are usually classified by the number of POLES and number of THROWS. Poles are shown in schematics as those contacts through which current enters the switch; they are connected to the movable contacts. Each pole may be connected to another part of the circuit through the switch by "throwing" the switch (movable contacts) to another position. This action provides an individual conduction path through the switch for each pole connection. The number of THROWS indicates the number of different circuits that can be controlled by each pole. By counting the number of points where current enters the switch (from the schematic symbol or the switch itself), you can determine the number of poles. By counting the number of different points each pole can connect with, you can determine the number of throws.

Figure 3-3 will help you understand this concept by showing illustrations of various multicontact switches and their schematic symbols.

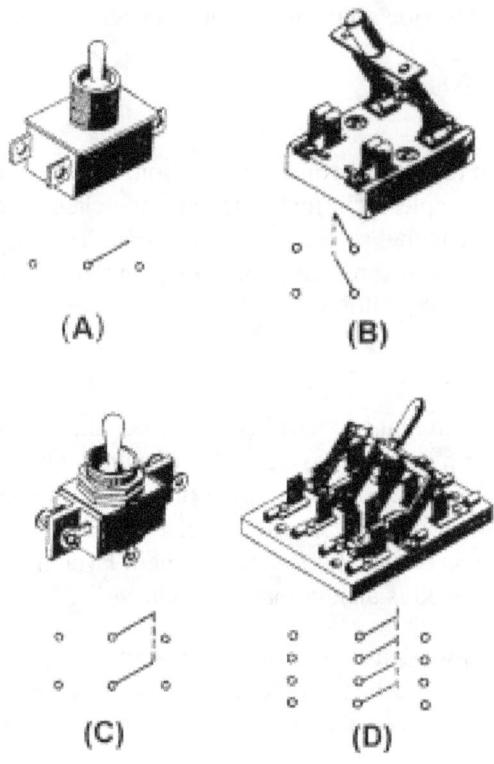

Figure 3-3.—Multicontact switches.

Figure 3-3(A) shows a single-pole, double-throw switch. The illustration shows three terminals (connections) on this switch. The schematic symbol for the switch is also shown.

The center connection of the schematic symbol represents the point at which current enters the switch. The left and right connections represent the two different points to which this current can be switched. From the schematic symbol, it is easy to determine that this is a single-pole, double-throw switch.

Now look at figure 3-3(B). The switch is shown with its schematic symbol. The schematic symbol has two points at which current can enter the switch, so this is a double-pole switch. Each of the poles is mechanically connected (still electrically separate) to one point, so this is a single-throw switch. Only one throw is required to route two separate circuit paths through the switch.

Figure 3-3(C) shows a double-pole, double-throw switch and its schematic symbol. Figure 3-3(D) shows a four-pole, double-throw switch and its schematic symbol.

It might help you to think of switches with more than one pole as several switches connected together mechanically. For example, the knife switch shown in figure 3-3(D) could be thought of as four single-pole, double-throw switches mechanically connected together.

Q4. What is the difference between a manual and an automatic switch?

Q5. What is one example of a manual switch?

Q6. What is one example of an automatic switch?

Q7. Why are multicontact switches used?

Q8. Label the schematic symbols shown in figure 3-4 as to number of poles and number of throws.

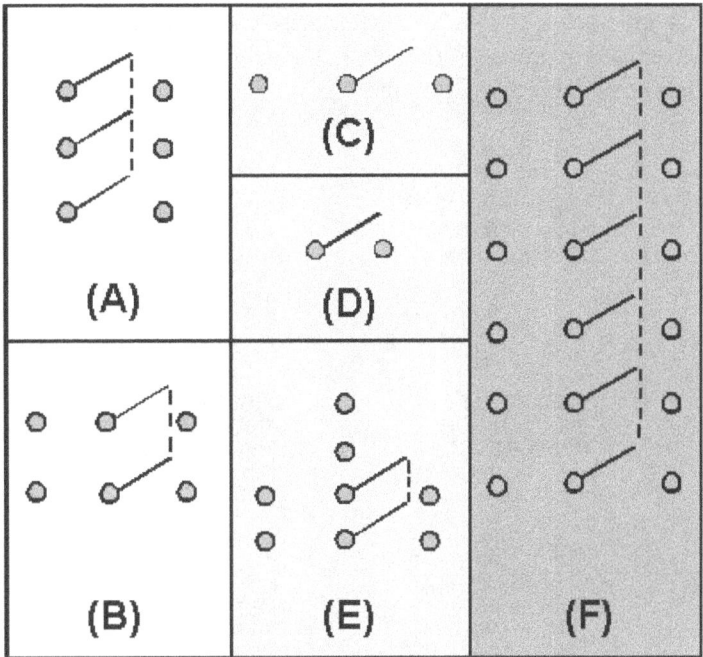

Figure 3-4.—Schematic symbols of switches.

Single-Break and Double-Break Switches

Switches can also be classified as SINGLE-BREAK or DOUBLE-BREAK switches. This refers to the number of places in which the switch opens or breaks the circuit. All of the switches shown so far have been single-break switches. A double-break switch is shown in figure 3-5. The schematic symbol shown in figure 3-5(A) shows that this switch breaks the circuit in two places (at both terminals). The upper part of the schematic symbol indicates that these contacts are in the open position and the circuit will close when the switch is acted upon (manually or automatically). The lower symbol shows closed contacts. These contacts will open the circuit when the switch is acted upon.

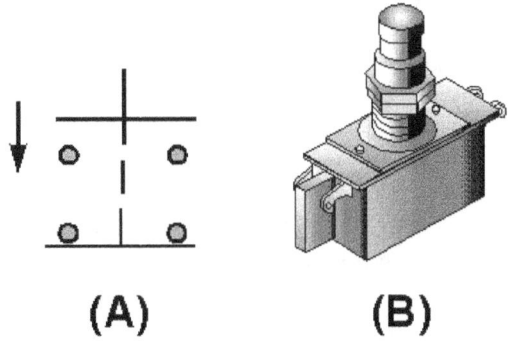

Figure 3-5.—Double-break pushbutton switch.

Figure 3-5(B) is a picture of the switch. This switch is called a pushbutton switch because it has a button that must be pushed to change the switch contact connections. Notice that the switch has four terminals. The schematic symbol in figure 3-5(A) shows that when one set of contacts is open, the other set of contacts is closed. This switch is a double-pole, single-throw, double-break switch.

The number of poles in a switch is independent of the number of throws and whether it is a single or double break switch. The number of throws in a switch is independent of the number of poles and whether it is a single or double break switch. In other words, each characteristic of a switch (poles, throws, break) is not determined by either of the other characteristics. Figure 3-6 shows the schematic symbols for several different switch configurations.

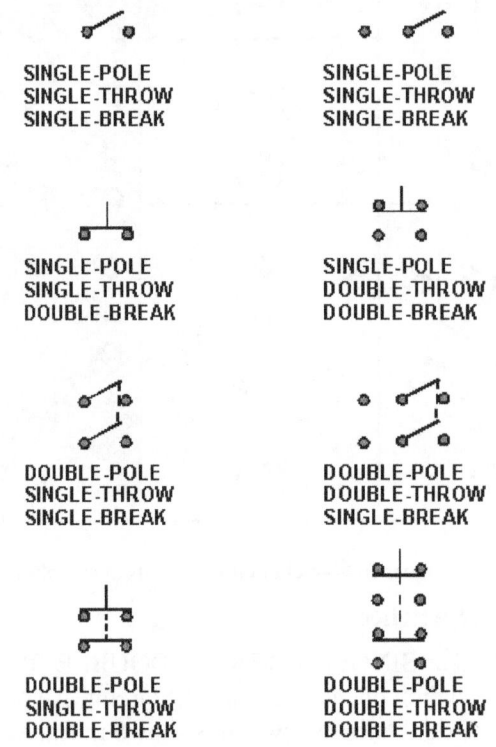

SINGLE-POLE
SINGLE-THROW
SINGLE-BREAK

SINGLE-POLE
SINGLE-THROW
SINGLE-BREAK

SINGLE-POLE
SINGLE-THROW
DOUBLE-BREAK

SINGLE-POLE
DOUBLE-THROW
DOUBLE-BREAK

DOUBLE-POLE
SINGLE-THROW
SINGLE-BREAK

DOUBLE-POLE
DOUBLE-THROW
SINGLE-BREAK

DOUBLE-POLE
SINGLE-THROW
DOUBLE-BREAK

DOUBLE-POLE
DOUBLE-THROW
DOUBLE-BREAK

Figure 3-6.—Schematic symbols of switch configurations.

Rotary Switches

A rotary switch is a midcontact switch part of the schematic with the contacts arranged in a full or partial circle. Instead of a pushbutton or toggle, the mechanism used to select the contact moves in a circular motion and must be turned. Rotary switches can be manual or automatic switches. An automobile distributor, the ignition switch on a motor vehicle, and the channel selector on some television sets are rotary switches.

The automobile distributor cap and rotor are an example of the simplest form of an automatic rotary switch. Figure 3-7 shows a portion of an automobile ignition system with the distributor cap and rotor shown. The rotor is the portion of this switch that moves (rotates) and selects the circuit (spark plug). The rotor does not actually touch the contacts going to the spark plugs, but the signal (spark) jumps the gap between the rotor and the contacts. This switch has one input (the rotor) and six positions (one for each spark plug). The schematic diagram for this rotary switch is shown below the illustration of the distributor cap.

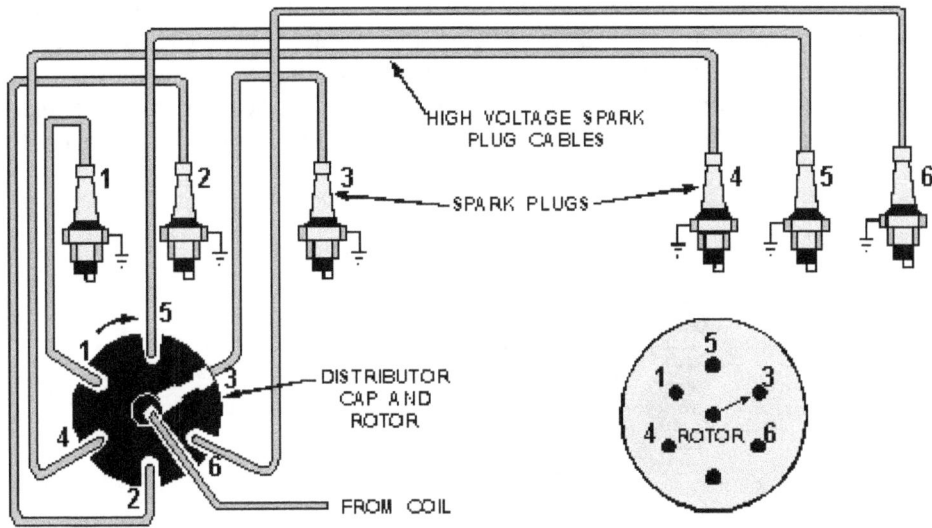

Figure 3-7.—Rotary switch in automobile ignition system.

The rotor in the distributor rotates continually (when in use) in one direction and makes a complete circle. This is not true for all rotary switches. The ignition switch in an automobile is also a rotary switch. It usually has four positions (accessory, off, on, start). Unlike the rotor, it does not rotate continually when in use, can be turned in either direction, and does not move through a complete circle.

Some rotary switches are made with several layers or levels. The arrangement makes possible the control of several circuits with a single switch. Figure 3-8 is an illustration of a rotary switch with two layers. Each layer has a selector and 20 contacts. As this switch is rotated, both layers select a single circuit (contact) of the 20.

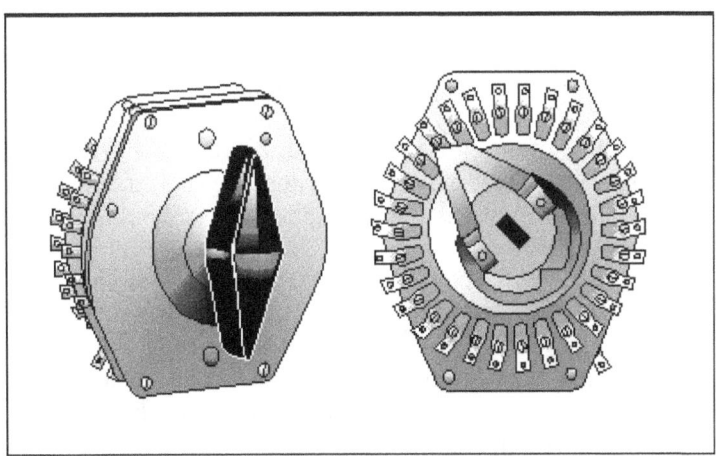

Figure 3-8.—Two-layer rotary switch.

The channel selector on some television sets is a multilayer rotary switch. It is also called a WAFER SWITCH. In a wafer switch, each layer is known as a wafer.

The schematic of the wafer is always drawn to represent the wafer as it would look if viewed from opposite the operating handle or mechanism. If the wafer has contacts on both sides, two drawings are used to show the two sides of the wafer. The two drawings are labeled "front" and "rear." The drawing labeled "front" represents the side of the wafer closest to the operating mechanism.

Figure 3-9(A) shows one wafer of a wafer switch and its schematic symbol. Contact 1 is the point at which current enters the wafer. It is always connected to the movable portion of the wafer. With the wafer in the position shown, contact 1 is connected to both contact 5 and 6 through the movable portion. If the movable portion was rotated slightly clockwise, contact 1 would only be connected to contact 5. This arrangement is known as MAKE BEFORE BREAK because the switch makes a contact before breaking the old contact.

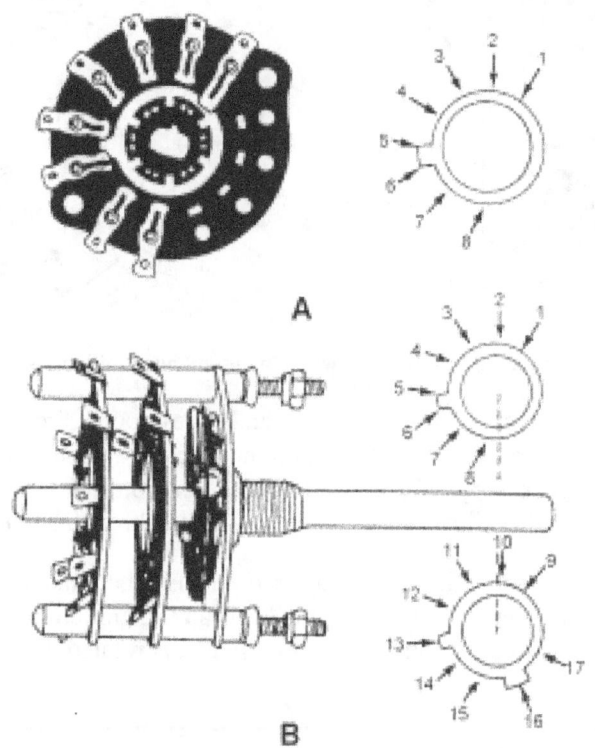

Figure 3-9.—Wafer switch.

Figure 3-9(B) is an illustration of the entire switch and its schematic symbol. Since the switch has two wafers mechanically connected by the shaft of the switch, the shaft rotates the movable portion of both wafers at the same time. This is represented on the schematic symbol by the dotted line connecting the two wafers.

The upper wafer of the schematic symbol is the wafer closest to the control mechanism, and is identical to the wafer shown in figure 3-9(A). When switches have more than one wafer, the first wafer shown is always the wafer closest to the operating mechanism. The lower wafer on the schematic diagram is the wafer farthest away from the operating mechanism. Contact 9 of this wafer is connected to the movable portion and is the point at which current enters the wafer. In the position shown, contact 9 is connected to both contact 13 and 16. If the switch is rotated slightly clockwise, contact 9 would no longer be connected to contact 13. A further clockwise movement would connect contact 9 to contact 12. This arrangement is called BREAK BEFORE MAKE. Contact 9 will also be connected to contact 15 at the same time as it is connected to contact 12.

Q9. Label the switch schematics shown in figure 3-10A through 3-10G.

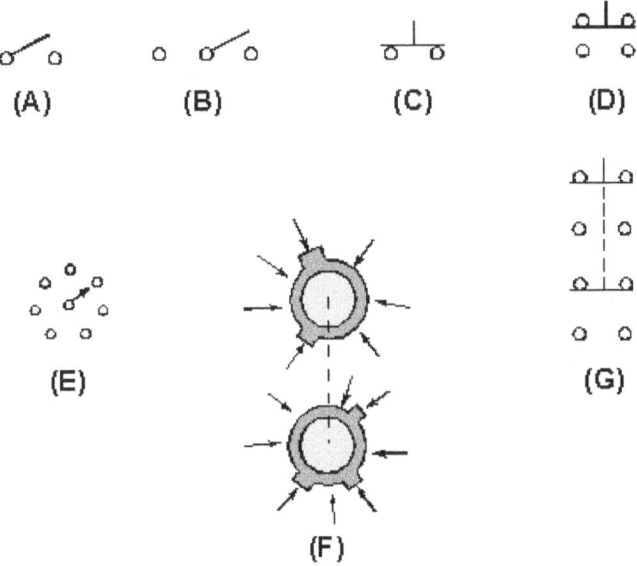

Figure 3-10.—Switch schematic symbols.

OTHER TYPES OF SWITCHES

You have learned that switches are classified by the number of poles, throws, and breaks. There are other factors used to describe a switch such as the type of actuator and the number of positions. In addition, switches are classified by whether the switch has momentary contacts or is locked into or out of position and whether or not the switch is snap-acting.

Type of Actuator

In addition to the pushbutton, toggle, and knife actuated switches already described, switches can have other actuators. There are rocker switches, paddle switches, keyboard switches and mercury switches (in which a small amount of mercury makes the electrical contact between two conductors).

Number of Positions

Switches are also classified by the number of positions of the actuating device. Figure 3-11 shows three toggle switches, the toggle positions, and schematic diagrams of the switch. Figure 3-11(A) is a single-pole, single-throw, two-position switch. The switch is marked to indicate the ON position (when the switch is closed) and the OFF position (when the switch is open). Figure 3-11(B) is a single-pole, double-throw, three-position switch. The switch markings show two ON positions and an OFF position. When this switch is OFF, no connection is made between any of the terminals. In either of the ON positions, the center terminal is connected to one of the outside terminals. (The outside terminals are not connected together in any position of the switch.) Figure 3-11(C) is a single-pole, double-throw, two-position switch. There is no OFF position. In either position of this switch, the center terminal is connected to one of the outside terminals.

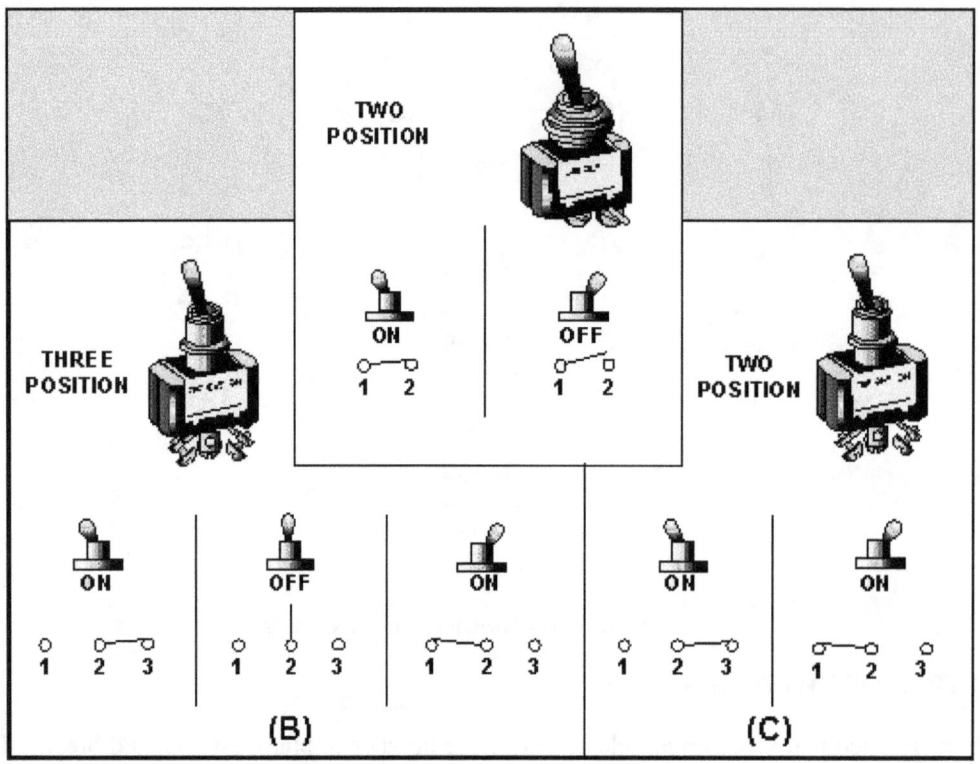

Figure 3-11.—Two- and three-position switches.

Momentary and Locked Position Switches

In some switches, one or more of the switch positions are MOMENTARY. This means that the switch will only remain in the momentary position as long as the actuator is held in that position. As soon as you let go of the actuator, the switch will return to a non-momentary position. The starter switch on an automobile is an example of a momentary switch. As soon as you release the switch, it no longer applies power to the starter.

Another type of switch can be LOCKED IN or OUT of some of the switch positions. This locking prevents the accidental movement of the switch. If a switch has locked-in positions, the switch cannot be moved from those positions accidentally (by the switch being bumped or mistaken for an unlocked switch). If the switch has locked-out positions, the switch cannot be moved into those positions accidentally. Figure 3-12 shows a three-position, locking switch.

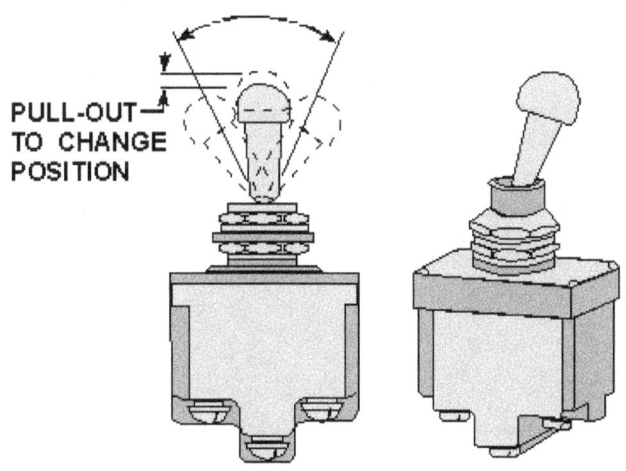

Figure 3-12.—Three-position locking switch.

Snap-Acting Switches

A SNAP-ACTING switch is a switch in which the movement of the switch mechanism (contacts) is relatively independent of the activating mechanism movement. In other words, in a toggle switch, no matter how fast or slow you move the toggle, the actual switching of the circuit takes place at a fixed speed. The snap-acting switch is constructed by making the switch mechanism a leaf spring so that it "snaps" between positions. A snap-acting switch will always be in one of the positions designed for that switch. The switch cannot be "between" positions. A two-position, single-pole, double-throw, snap-acting switch could not be left in an OFF position.

Accurate Snap-Acting Switches

An ACCURATE SNAP-ACTING SWITCH is a snap-acting switch in which the operating point is pre-set and very accurately known. The operating point is the point at which the plunger causes the switch to "switch." The accurate snap-acting switch is commonly called a MICROSWITCH. A microswitch is shown in figure 3-13.

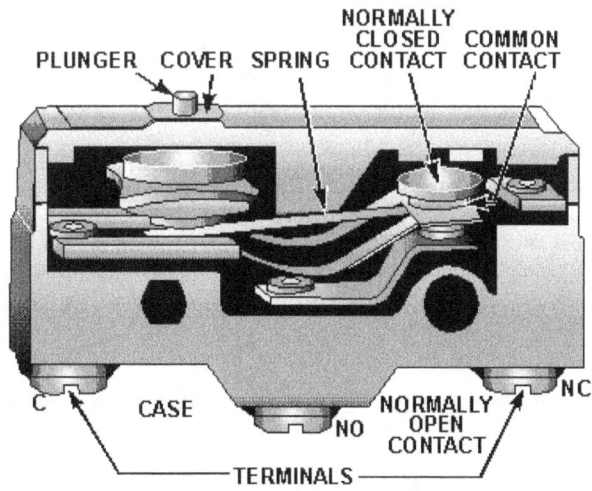

Figure 3-13.—Accurate snap-acting switch (microswitch).

The full description of the microswitch shown in figure 3-13 is a two-position, single-pole, double-throw, single-break, momentary-contact, accurate, snap-acting switch. Notice the terminals marked C, NO, and NC. These letters stand for common, normally open, and normally closed. The common terminal is connected to the normally closed terminal until the plunger is depressed. When the plunger is depressed, the spring will "snap" into the momentary position and the common terminal will be connected to the normally open terminal. As soon as the plunger is released, the spring will "snap" back to the original condition.

This basic accurate snap-acting switch is used in many applications as an automatic switch. Several different methods are used to actuate this type of switch. Some of the more common actuators and their uses are shown in figure 3-14.

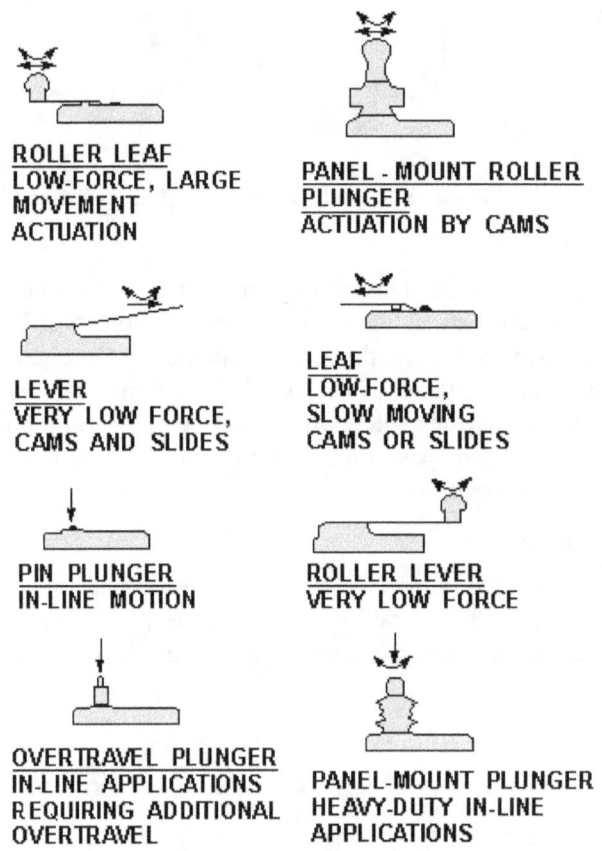

Figure 3-14.—Common actuators and their uses for accurate snap-acting switches.

Q10. What classification of a switch is used when you describe it as a rocker switch?

Q11. In describing a switch by the number of positions of the actuator, what are the two possible configurations for a single-pole, double-throw switch?

Q12. What type of switch should be used to control a circuit that requires a temporary actuation signal?

Q13. What type of switch is used if it is necessary to guard against a circuit being accidentally turned on or off?

Q14. What is the common name used for an accurate snap-acting switch?

SWITCH RATING

Switches are rated according to their electrical characteristics. The rating of a switch is determined by such factors as contact size, contact material, and contact spacing. There are two basic parts to a switch rating-the current and voltage rating. For example, a switch may be rated at 250 volts dc, 10 amperes. Some switches have more than one rating. For example, a single switch may be rated at 250 volts dc, 10 amperes; 500 volts ac, 10 amperes; and 28 volts dc, 20 amperes. This rating indicates a current rating that depends upon the voltage applied.

CURRENT RATING OF A SWITCH

The current rating of a switch refers to the <u>maximum</u> current the switch is designed to carry. This rating is dependent on the voltage of the circuit in which the switch is used. This is shown in the example given above. The current rating of a switch should never be exceeded. If the current rating of a switch is exceeded, the contacts may "weld" together making it impossible to open the circuit.

VOLTAGE RATING OF A SWITCH

The voltage rating of a switch refers to the <u>maximum</u> voltage allowable in the circuit in which the switch is used. The voltage rating may be given as an ac voltage, a dc voltage, or both. The voltage rating of a switch should never be exceeded. If a voltage higher than the voltage rating of the switch is applied to the switch, the voltage may be able to "jump" the open contacts of the switch. This would make it impossible to control the circuit in which the switch was used.

Q15. What is the current rating of a switch?

Q16. What is the voltage rating of a switch?

MAINTENANCE AND REPLACEMENT OF SWITCHES

Switches are usually a very reliable electrical component. This means, they don't fail very often. Most switches are designed to operate 100,000 times or more without failure if the voltage and current ratings are not exceeded. Even so, switches do fail. The following information will help you in maintaining and changing switches.

CHECKING SWITCHES

There are two basic methods used to check a switch. You can use an ohmmeter or a voltmeter. Each of these methods will be explained using a single-pole, double-throw, single-break, three-position, snap-acting, toggle switch.

Figure 3-15 is used to explain the method of using an ohmmeter to check a switch. Figure 3-15(A) shows the toggle positions and schematic diagrams for the three switch positions. Figure 3-15(B) shows the ohmmeter connections used to check the switch while the toggle is in position 1. Figure 3-15(C) is a table showing the switch position, ohmmeter connection, and correct ohmmeter reading for those conditions.

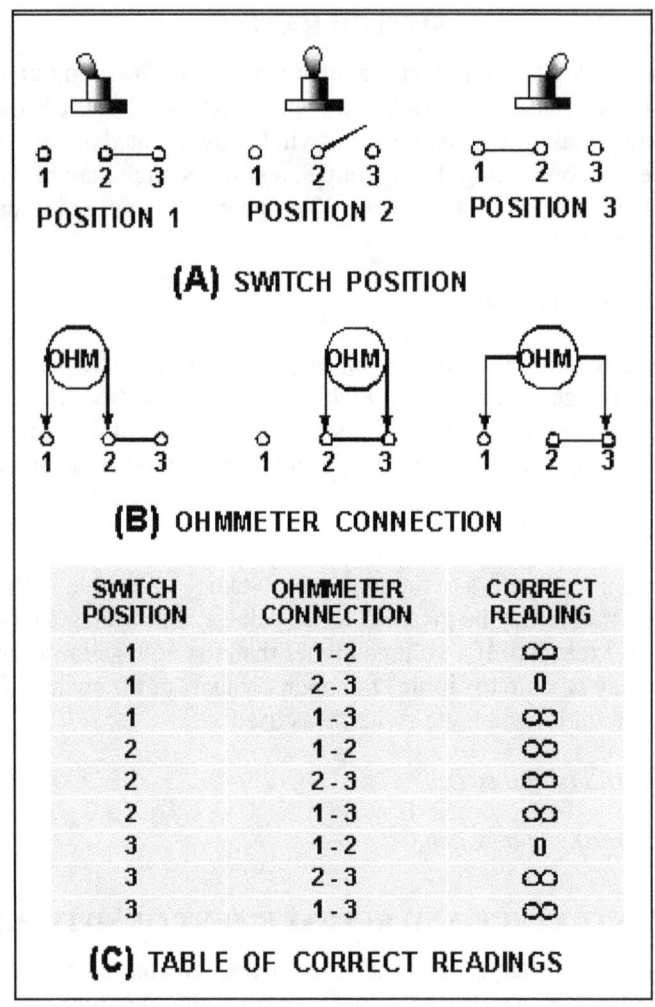

Figure 3-15.—Table of correct readings.

SWITCH POSITION	OHMMETER CONNECTION	CORRECT READING
1	1 - 2	∞
1	2 - 3	0
1	1 - 3	∞
2	1 - 2	∞
2	2 - 3	∞
2	1 - 3	∞
3	1 - 2	0
3	2 - 3	∞
3	1 - 3	∞

(C) TABLE OF CORRECT READINGS

With the switch in position 1 and the ohmmeter connected to terminals 1 and 2 of the switch, the ohmmeter should indicate (∞). When the ohmmeter is moved to terminals 2 and 3, the ohmmeter should indicate zero ohms. With the ohmmeter connected to terminals 1 and 3, the indication should be (∞).

As you remember from chapter 1, before the ohmmeter is used, power must be removed from the circuit and the component being checked should be isolated from the circuit. The best way to isolate the switch is to remove it from the circuit completely. This is not always practical, and it is sometimes necessary to check a switch while there is power applied to it. In these cases, you would not be able to use an ohmmeter to check the switch, but you can check the switch by the use of a voltmeter.

Figure 3-16(A) shows a switch connected between a power source (battery) and two loads. In figure 3-16(B), a voltmeter is shown connected between ground and each of the three switch terminals while the switch is in position 1. Figure 3-16(C) is a table showing the switch position, voltmeter connection, and the correct voltmeter reading.

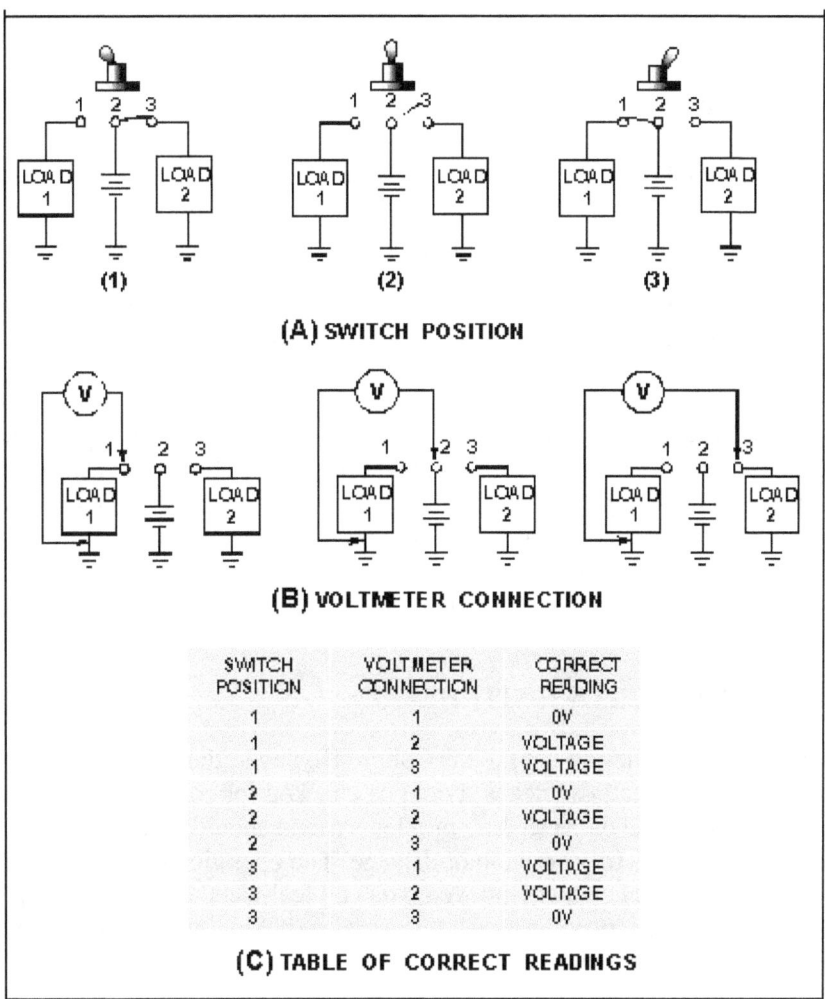

Figure 3-16.—Table of correct readings.

SWITCH POSITION	VOLTMETER CONNECTION	CORRECT READING
1	1	0V
1	2	VOLTAGE
1	3	VOLTAGE
2	1	0V
2	2	VOLTAGE
2	3	0V
3	1	VOLTAGE
3	2	VOLTAGE
3	3	0V

(C) TABLE OF CORRECT READINGS

With the switch in position 1 and the voltmeter connected between ground and terminal 1, the voltmeter should indicate no voltage (OV). When the voltmeter is connected to terminal 2, the voltmeter should indicate the source voltage. With the voltmeter connected to terminal 3, the source voltage should also be indicated. The table in figure 3-16(C) will show you the correct readings with the switch in position 2 or 3.

REPLACEMENT OF SWITCHES

When a switch is faulty, it must be replaced. The technical manual for the equipment will specify the exact replacement switch. If it is necessary to use a substitute switch, the following guidelines should be used. The substitute switch must have all of the following characteristics.

- At least the same number of poles.

- At least the same number of throws.

- The same number of breaks.

- At least the same number of positions.

- The same configuration in regard to momentary or locked positions.

- A voltage rating equal to or higher than the original switch.

- A current rating equal to or higher than the original switch.

- A physical size compatible with the mounting.

In addition, the type of actuator (toggle, pushbutton, rocker, etc.) should be the same as the original switch. (This is desirable but not necessary. For example, a toggle switch could be used to replace a rocker switch if it were acceptable in all other ways.)

The number of poles and throws of a switch can be determined from markings on the switch itself. The switch case will be marked with a schematic diagram of the switch or letters such as SPST for single-pole, single-throw. The voltage and current ratings will also be marked on the switch. The number of breaks can be determined from the schematic marked on the switch or by counting the terminals after you have determined the number of poles and throws. The type of actuator, number of positions, the momentary and locked positions of the switch can all be determined by looking at the switch and switching it to all the positions.

PREVENTIVE MAINTENANCE OF SWITCHES

As already mentioned, switches do not fail very often. However, there is a need for preventive maintenance of switches. Periodically switches should be checked for corrosion at the terminals, smooth and correct operation, and physical damage. Any problems found should be corrected immediately. Most switches can be inspected visually for corrosion or damage. The operation of the switch may be checked by moving the actuator. When the actuator is moved, you can feel whether the switch operation is smooth or seems to have a great deal of friction. To check the actual switching, you can observe the operation of the equipment or check the switch with a meter.

Q17. What two types of meters can be used to check a switch?

Q18. If a switch must be checked with power applied, what type of meter is used?

Q19. A double-pole, double-throw, single-break, three-position, toggle switch is faulty. This switch has a momentary position 1 and is locked out opposition 3. The voltage and current ratings for the switch are 115 volt dc, 5 amperes. No direct replacement is available. From switches A through I, in table 3-1, indicate if the switch is acceptable or not acceptable as a substitute. Of the acceptable switches, rank them in order of choice. If the switch is unacceptable, give the reason.

Q20. What should you check when performing preventive maintenance on a switch?

Table 3-1.—Replacement Switches and Their Characteristics

	POLES	THROWS	BREAKS	NUMBER OF POSITIONS	MOMENTARY POSITIONS	LOCKED POSITIONS	ACTUATOR	RATING
A	2	1	1	2	—	—	PUSH BUTTON	115Vdc 5A
B	2	2	2	3	1	OUT-3	TOGGLE	150Vdc 5A
C	2	2	1	3	1	OUT-3	ROCKER	115Vdc 10A
D	1	2	1	3	1	OUT-3	TOGGLE	115Vdc 5A
E	2	2	1	3	—	OUT-3	ROCKER	150Vdc 10A
F	2	2	1	3	1	OUT-3	TOGGLE	150Vdc 10A
G	2	2	1	3	1	IN-3	TOGGLE	115Vdc 10A
H	2	2	1	3	1	OUT-3	ROCKER	115Vdc 3A
I	2	2	1	3	1	OUT-3	ROCKER	28Vdc 5A

SOLENOIDS

A SOLENOID is a control device that uses electromagnetism to convert electrical energy into mechanical motion. The movement of the solenoid may be used to close a set of electrical contacts, cause the movement of a mechanical device, or both at the same time.

Figure 3-17 is a cutaway view of a solenoid showing the solenoid action. A solenoid is an electromagnet formed by a conductor wound in a series of loops in the shape of a spiral. Inserted within this coil is a soft-iron core and a movable plunger. The soft-iron core is pinned or held in an immovable position. The movable plunger (also soft iron) is held away from the core by a spring when the solenoid is deenergized.

When current flows through the conductor, it produces a magnetic field. The magnetic flux produced by the coil results in establishing north and south poles in both the core and the plunger. The plunger is attracted along the lines of force to a position at the center of the coil. As shown in figure 3-17, the deenergized position of the plunger is partially out of the coil due to the action of the spring. When voltage is applied, the current through the coil produces a magnetic field. This magnetic field draws the plunger within the coil, resulting in mechanical motion. When the coil is deenergized, the plunger returns to its normal position because of spring action. The effective strength of the magnetic field on the plunger varies according to the distance between the plunger and the core. For short distances, the strength of the field is strong; and as distances increase, the strength of the field drops off quite rapidly.

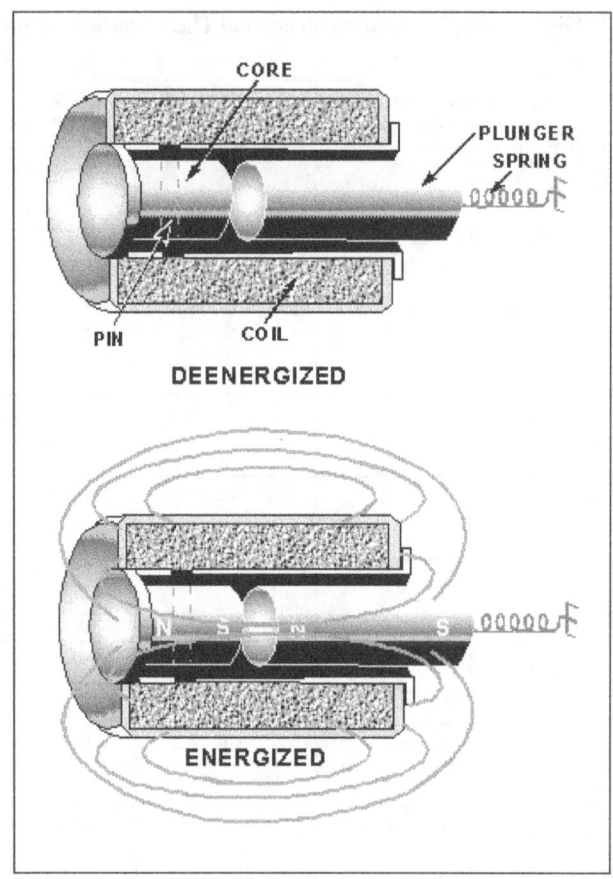

Figure 3-17.—Solenoid action.

While a solenoid is a control device, the solenoid itself is energized by some other control device such as a switch or a relay. One of the distinct advantages in the use of solenoids is that a mechanical movement can be accomplished at a considerable distance from the control device. The only link necessary between the control device and the solenoid is the electrical wiring for the coil current. The solenoid can have large contacts for the control of high current. Therefore, the solenoid also provides a means of controlling high current with a low current switch. For example, the ignition switch on an automobile controls the large current of a starter motor by the use of a solenoid. Figure 3-18 shows a cutaway view of a starter motor-solenoid combination and a section of the wiring for the solenoid. Notice that the solenoid provides all electrical contact for current to the starter motor as well as a mechanical movement of the shift lever.

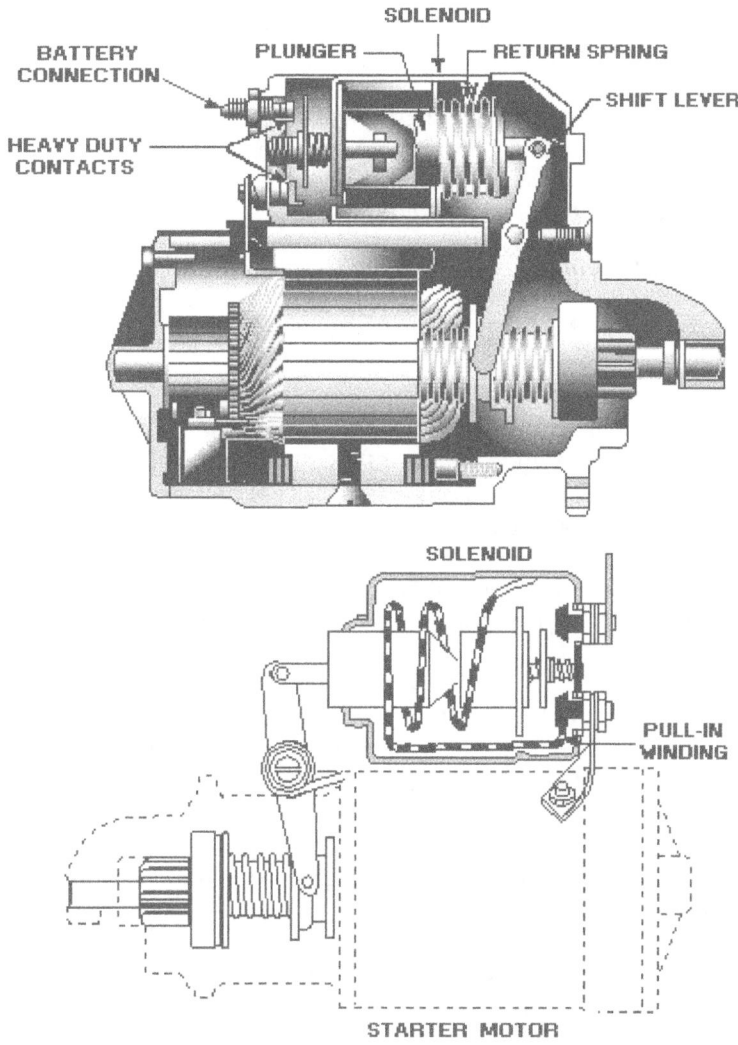

Figure 3-18.—Starter motor and solenoid.

MAINTENANCE OF SOLENOIDS

If you suspect that a solenoid is not working properly, the first step in troubleshooting it is a good visual inspection. Check the connections for poor soldering, loose connections, or broken wires. The plunger should be checked for cleanliness, binding, mechanical failure, and improper alignment. The mechanism that the solenoid is connected to (actuates) should also be checked for proper operation.

The second step is to check the energizing voltage with a voltmeter. If the voltage is too low, the result is less current flowing through the coil and a weak magnetic field. A weak magnetic field can result in slow or poor operation. Low voltage could also result in chatter or no operation at all. If the energizing voltage is too high, it could damage the solenoid by causing overheating or arcing. In either case, the voltage should be reset to the proper value so that further damage or failure of the solenoid will not result.

The solenoid coil should then be checked for opens, shorts, and proper resistance with an ohmmeter. If the solenoid coil is open, current cannot flow through it and the magnetic field is lost. A short results in fewer turns and higher current in the coil. The net result of a short is a weak magnetic field. A high-resistance coil will reduce coil current and also result in a weak magnetic field. A weak magnetic field will cause less attraction between the plunger and the core of the coil. This will result in improper operation similar to that caused by low voltage. If the coil is open, shorted, or has changed in resistance, the solenoid should be replaced.

Finally, you should check the solenoid to determine if the coil is shorted to ground. If a short to ground is found, the short should be removed to restore the solenoid to proper operation.

Q21. What is the operating principle of a solenoid?

Q22. What is one example of the use of a solenoid?

Q23. If a solenoid is not operating properly, what items should be checked?

RELAYS

The RELAY is a device that acts upon the same fundamental principle as the solenoid. The difference between a relay and a solenoid is that a relay does not have a movable core (plunger) while the solenoid does. Where multipole relays are used, several circuits may be controlled at once.

Relays are electrically operated control switches, and are classified according to their use as POWER RELAYS or CONTROL RELAYS. Power relays are called CONTACTORS; control relays are usually known simply as relays.

The function of a contactor is to use a relatively small amount of electrical power to control the switching of a large amount of power. The contactor permits you to control power at other locations in the equipment, and the heavy power cables need be run only through the power relay contacts.

Only lightweight control wires are connected from the control switches to the relay coil. Safety is also an important reason for using power relays, since high power circuits can be switched remotely without danger to the operator.

Control relays, as their name implies, are frequently used in the control of low power circuits or other relays, although they also have many other uses. In automatic relay circuits, a small electric signal may set off a chain reaction of successively acting relays, which then perform various functions.

In general, a relay consists of a magnetic core and its associated coil, contacts, springs, armature, and the mounting. Figure 3-19 illustrates the construction of a relay. When the coil is energized, the flow of current through the coil creates a strong magnetic field which pulls the armature downward to contact C1, completing the circuit from the common terminal to C1. At the same time, the circuit to contact C2, is opened.

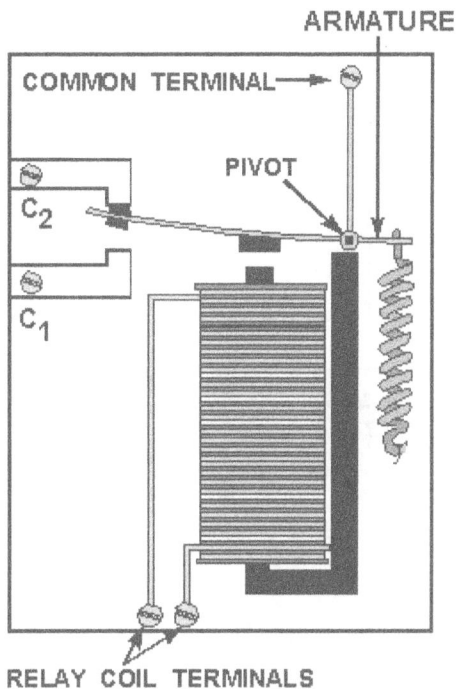

Figure 3-19.—Relay construction.

A relay can have many different types of contacts. The relay shown in figure 3-19 has contacts known as "break-make" contacts because they break one circuit and make another when the relay is energized. Figure 3-20 shows five different combinations of relay contacts and the names given to each.

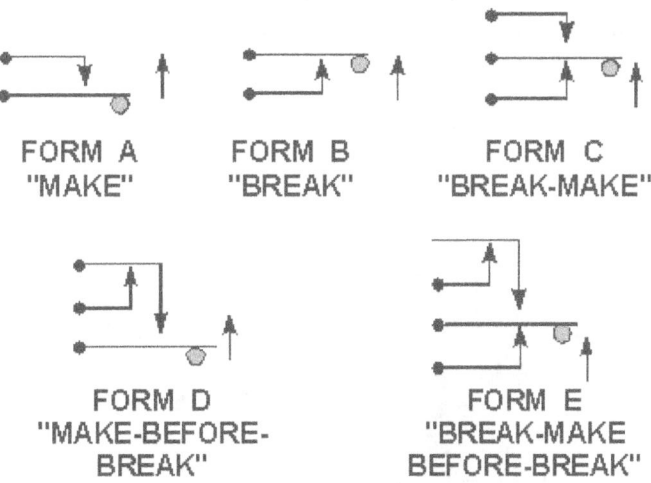

FORM A
"MAKE"

FORM B
"BREAK"

FORM C
"BREAK-MAKE"

FORM D
"MAKE-BEFORE-
BREAK"

FORM E
"BREAK-MAKE
BEFORE-BREAK"

Figure 3-20.—Contact combinations.

A single relay can have several different types of contact combinations. Figure 3-21 is the contact arrangement on a single relay that has four different contact combinations. (The letters next to the contacts are the "forms" shown in figure 3-20.)

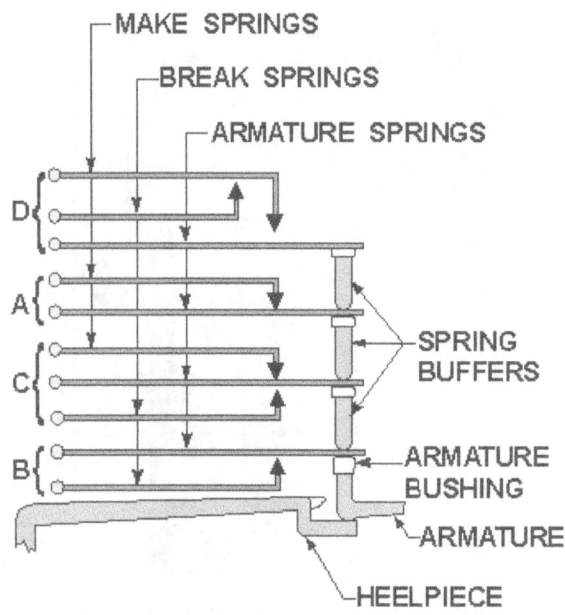

Figure 3-21.—Relay contact arrangement.

One type of relay with multiple sets of contacts is the clapper relay shown in figure 3-22. As the circuit is energized, the clapper is pulled to the magnetic coil. This physical movement of the armature of the clapper forces the pushrod and movable contacts upward. Any number of sets of contacts may be built onto the relay; thus, it is possible to control many different circuits at the same time. This type of relay can be a source of trouble because the motion of the clapper armature does not necessarily assure movement of all the movable contacts. Referring to figure 3-22, if the pushrod were broken, the clapper armature might push the lower movable contact upward but not move the upper movable contact.

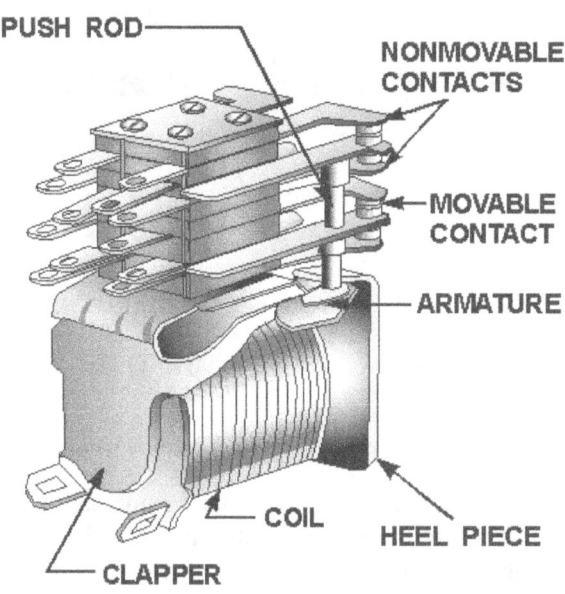

Figure 3-22.—Clapper-type relay.

Some equipment requires a "warm-up" period between the application of power and some other action. For example, vacuum tubes (covered later in this training series) require a delay between the application of filament power and high voltage. A time-delay relay will provide this required delay.

A thermal time-delay relay (fig. 3-23) is constructed to produce a delayed action when energized. Its operation depends on the thermal action of a bimetallic element similar to that used in a thermal circuit breaker. A heater is mounted around or near the element. The movable contact is mounted on the element itself. As the heat causes the element to bend (because of the different thermal expansion rates), the contacts close.

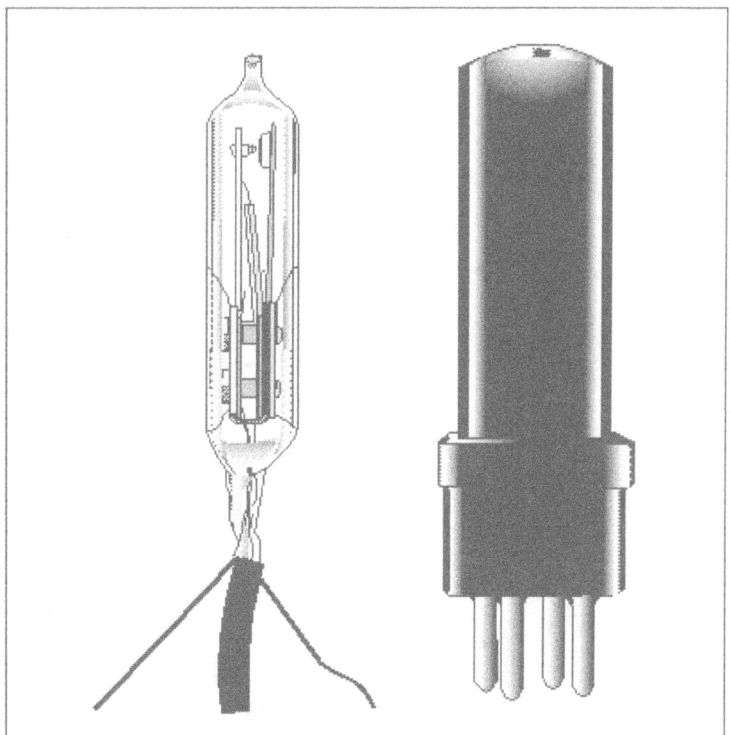

Figure 3-23.—A thermal time-delay relay.

Relays can be described by the method of packaging; open, semisealed, and sealed. Figure 3-24 shows several different relays and illustrates these three types of packaging.

Figure 3-24 (E), (G) and (H) are open relays. The mechanical motion of the contacts can be observed and the relays are easily available for maintenance. Figure 3-24 (A), (B) and (C) are semisealed relays. The covers provide protection from dust, moisture, and other foreign material but can be removed for maintenance.

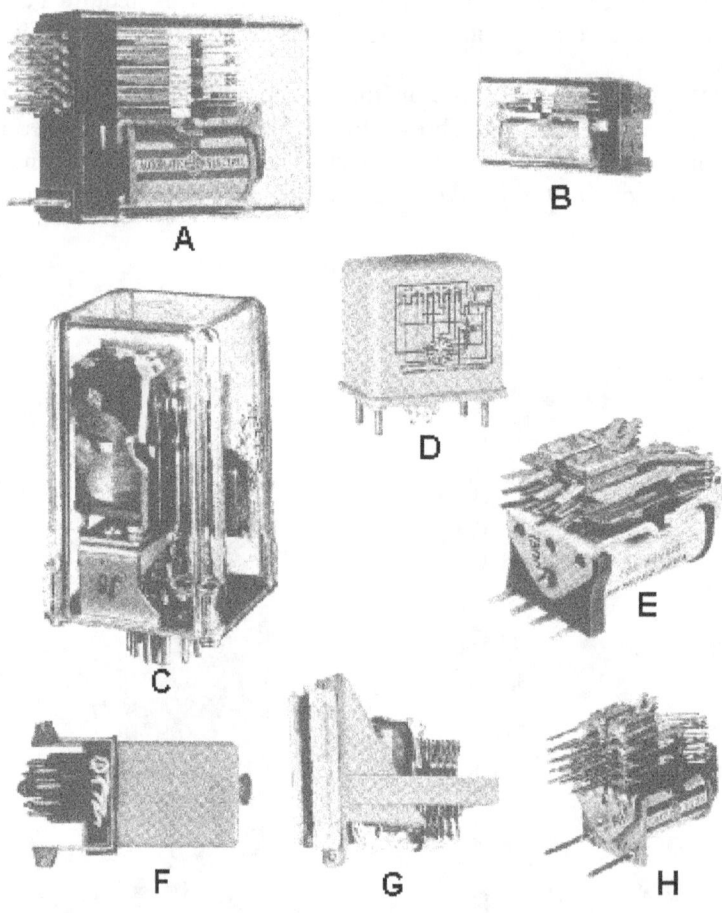

Figure 3-24.—Relay enclosures.

The clear plastic or glass covers provide a means of observing the operation of the relay without removal of the cover. Figure 3-24 (D) and (F) are examples of a hermetically sealed relay. These relays are protected from temperature or humidity changes as well as dust and other foreign material. Since the covers cannot be removed, the relays are also considered to be tamper-proof. With metal or other opaque covers, the operation of the relay can be "felt" by placing your finger on the cover and activating the relay.

Q24. What is the operating principle of a relay?

Q25. How does a relay differ from a solenoid?

Q26. What are the two classifications of relays?

MAINTENANCE OF RELAYS

The relay is one of the most dependable electromechanical devices in use, but like any other mechanical or electrical device, relays occasionally wear out or become inoperative. Should an inspection determine that a relay is defective, the relay should be removed immediately and replaced with another of the same type. You should be sure to obtain the same type relay as a replacement. Relays are rated in voltage, amperage, type of service, number of contacts, and similar characteristics.

Relay coils usually consist of a single coil. If a relay fails to operate, the coil should be tested for open circuit, short circuit, or short to ground. An open coil is a common cause of relay failure.

During preventive maintenance you should check for charred or burned insulation on the relay and for darkened or charred terminal leads. Both of these indicate overheating, and the likelihood of relay breakdown. One possible cause for overheating is that the power terminal connectors are not tight. This would allow arcing at the connection.

The build-up of film on the contact surfaces of a relay is another cause of relay trouble. Although film will form on the contacts by the action of atmospheric and other gases, grease film is responsible for a lot of contact trouble. Carbon build-up which is caused by the burning of a grease film or other substance (during arcing), also can be troublesome. Carbon forms rings on the contact surfaces and as the carbon rings build-up, the relay contacts are held open.

When current flows in one direction through a relay, a problem called "cone and crater" may be created at the contacts. The crater is formed by transfer of metal from one contact to the other contact, the deposit being in the shape of a cone. This condition is shown in figure 3-25(A).

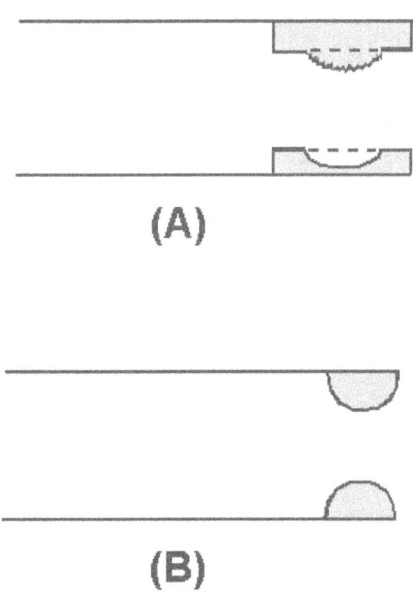

(A)

(B)

Figure 3-25.—Relay contacts.

Some relays are equipped with ball-shaped contacts and, in many applications, this type of contact is considered superior to a flat surface. Figure 3-25(B) shows a set of ball-shaped contacts. Dust or other substances are not as readily deposited on a ball-shaped surface. In addition, a ball-shaped contact penetrates film more easily than a flat contact. When you clean or service ball-shaped relay contacts, be careful to avoid flattening or otherwise altering the rounded surfaces of the contacts, YOU could damage a relay if you used sandpaper or emery cloth to clean the contacts. Only a burnishing tool, shown in figure 3-26 should be used for this purpose.

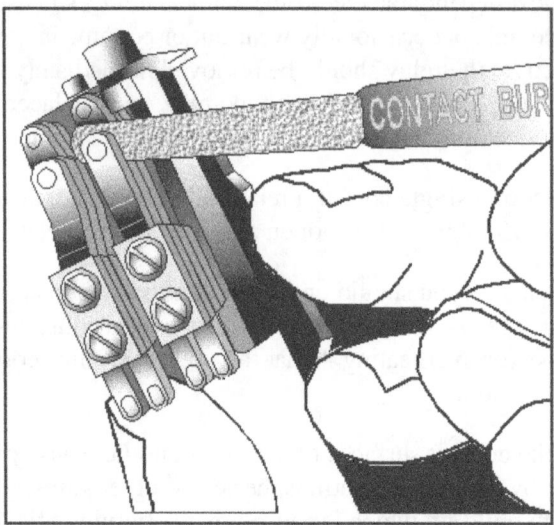

Figure 3-26.—Burnishing tool.

You should not touch the surfaces of the burnishing tool that are used to clean the relay contacts. After the burnishing, tool is used, it should be cleaned with alcohol.

Contact clearances or gap settings must be maintained in accordance with the operational specifications of the relay. When a relay has bent contacts, you should use a point bender (shown in figure 3-27) to straighten the contacts. The use of any other tool could cause further damage and the entire relay would have to be replaced.

Figure 3-27.—Point bender.

Cleanliness must be emphasized in the removal and replacement of covers on semi sealed relays. The entry of dust or other foreign material can cause poor contact connection. When the relay is installed in a position where there is a possibility of contact with explosive fumes, extra care should be taken with the cover gasket. Any damage to, or incorrect seating of the gasket increases the possibility of igniting the vapors.

Q27. How can you determine if a relay is operating (changing from one position to the other)?

Q28. What items should be checked on a relay that is not operating properly?

Q29. What is used to clean the contacts of a relay?

Q30. What tool is used to set contact clearances on a relay?

SUMMARY

This chapter has provided you with basic information on circuit control devices. The following is a summary of the main points in this chapter.

CIRCUIT CONTROL DEVICES are used to apply or remove power and to select a function or circuit within a device.

A **SWITCH** is one type of circuit control device. Switches are classified in many different ways.

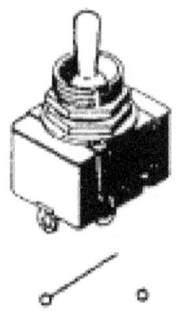

TOGGLE SWITCH

A **MANUAL SWITCH** must be tuned ON or OFF by a person. An AUTOMATIC SWITCH will turn a circuit ON or OFF without the action of a person by using mechanical or electrical devices.

MULTICONTACT SWITCHES make possible the control of more than one circuit or the selection of one of several possible circuits with a single switch.

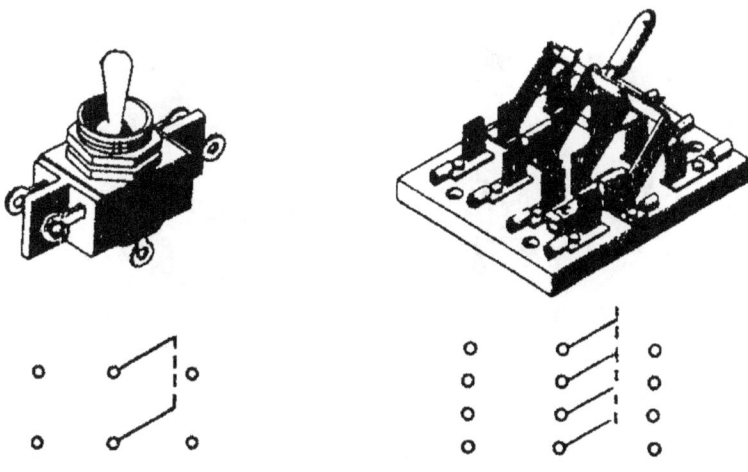

The **POLES** of a switch are the points at which current can enter the switch. The number of THROWS is the number of possible circuits that can be connected to each pole. The number of BREAKS is the number of points at which the switch breaks the circuit.

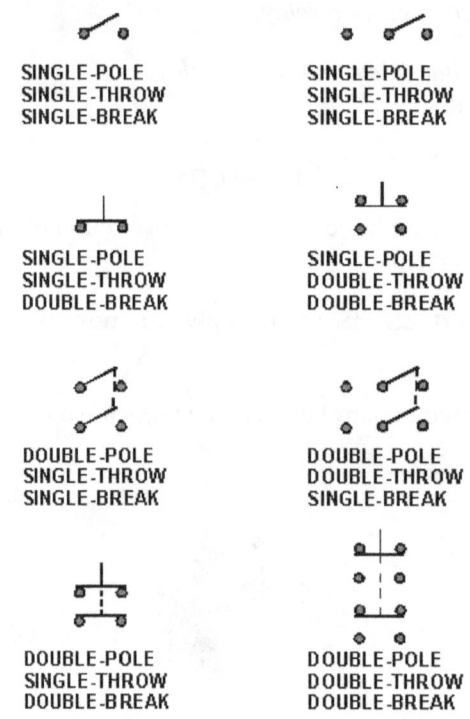

SINGLE-POLE
SINGLE-THROW
SINGLE-BREAK

SINGLE-POLE
SINGLE-THROW
SINGLE-BREAK

SINGLE-POLE
SINGLE-THROW
DOUBLE-BREAK

SINGLE-POLE
DOUBLE-THROW
DOUBLE-BREAK

DOUBLE-POLE
SINGLE-THROW
SINGLE-BREAK

DOUBLE-POLE
DOUBLE-THROW
SINGLE-BREAK

DOUBLE-POLE
SINGLE-THROW
DOUBLE-BREAK

DOUBLE-POLE
DOUBLE-THROW
DOUBLE-BREAK

A **ROTARY SWITCH** is a multicontact switch with contacts arranged in a circular or semicircular manner.

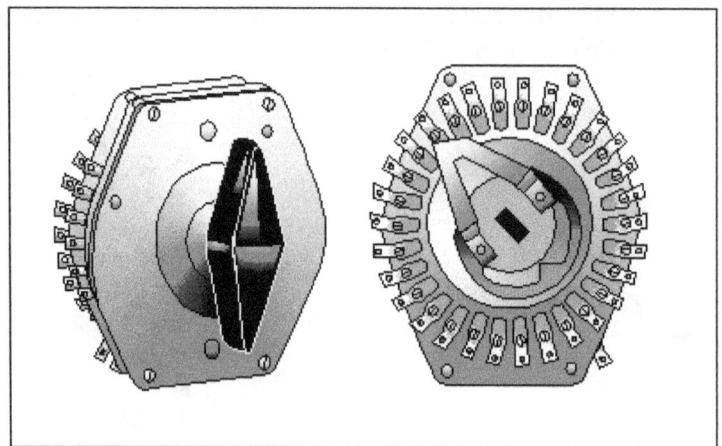

A **WAFER SWITCH** is a rotary switch in which the contacts are on wafers. The wafers are mechanically connected by the shaft of the switch.

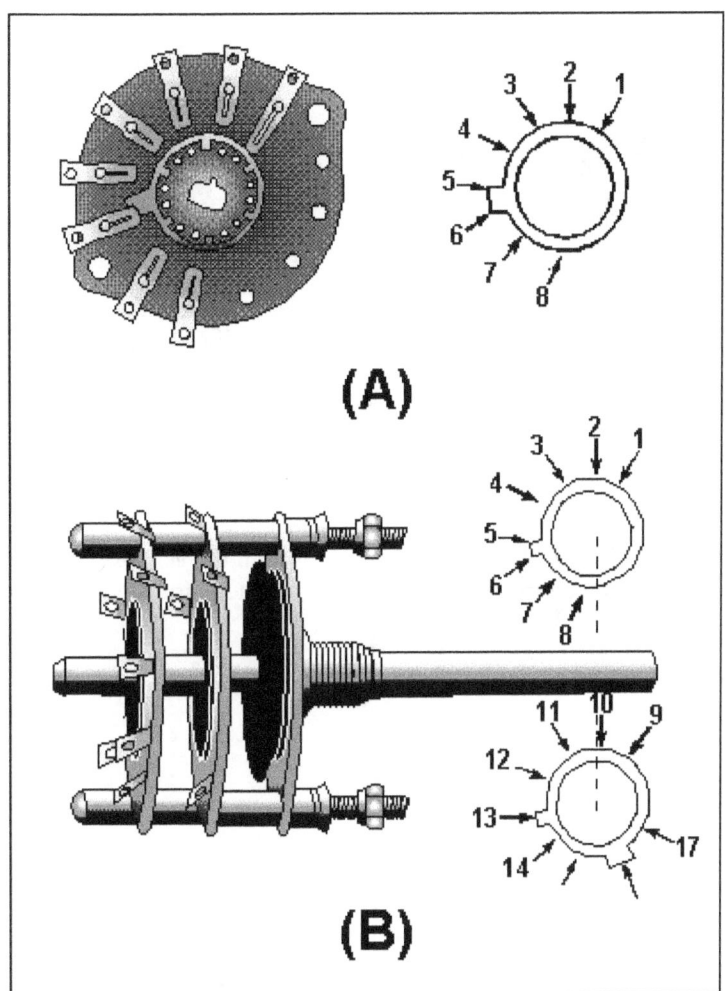

(A)

(B)

The **ACTUATOR** of a switch is the portion of the switch which is moved to cause the switch to change contact positions. The actuator could be a toggle, a pushbutton, a rocker, or, in the case of a rotary switch, a shaft and handle.

The **NUMBER OF POSITIONS** of a switch refers to the number of points at which the actuator can select a contact configuration.

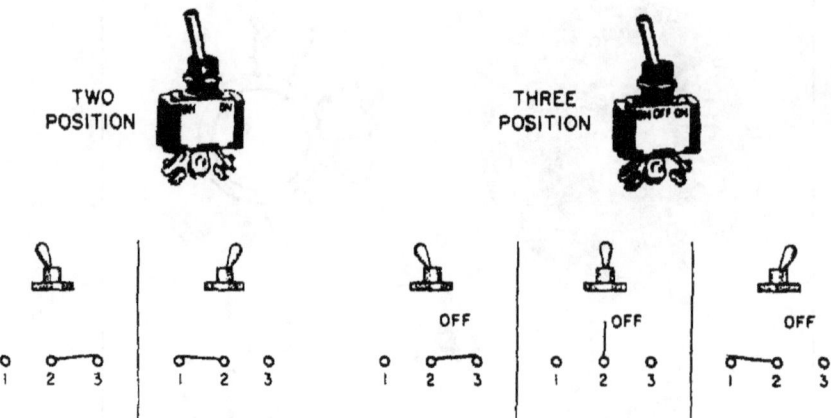

A **MOMENTARY POSITION** of a switch is one in which the actuator will only stay as long as force is applied to the actuator. When the force is removed, the actuator (and switch) will return to a non-momentary position.

A **LOCKED POSITION** of a switch is used to prevent the accidental movement of the actuator to or from a specific position.

PULL-OUT
TO CHANGE
POSITION

A **SNAP-ACTING SWITCH** is one in which the movement of the switch contacts is relatively independent of the actuator movement. This is accomplished by using a leaf spring for the common contact of the switch.

A **MICROSWITCH** is an accurate snap-acting switch and the operating point is preset and very accurately known.

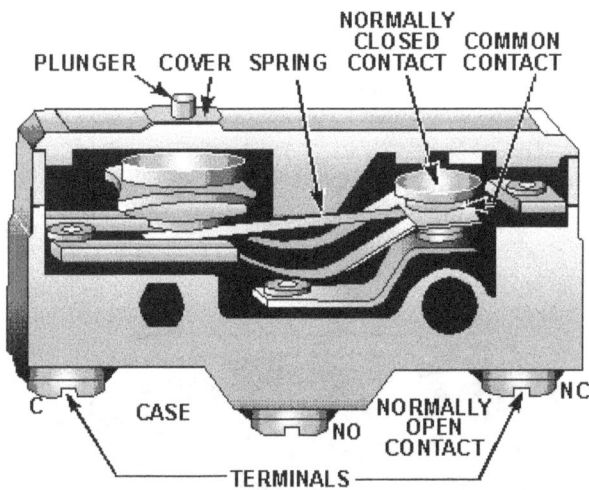

The **VOLTAGE RATING** of a switch is the maximum voltage the switch is designed to control. A voltage higher than the voltage rating may be able to "jump" the open contacts of the switch.

The **CURRENT RATING** of a switch is the maximum current the switch is designed to carry; it is dependent on the voltage rating. Any current higher than the current rating may cause the contacts of the switch to melt and "weld" together.

The contacts of a switch can be checked with an ohmmeter if power is removed or with a voltmeter if power is applied to the switch. To check a switch, the actuator should be checked for smooth and correct operation, the terminals should be checked for evidence of corrosion, and the physical condition of the switch should be determined. If a substitute switch must be used to replace a faulty switch, the substitute must have all of the following:

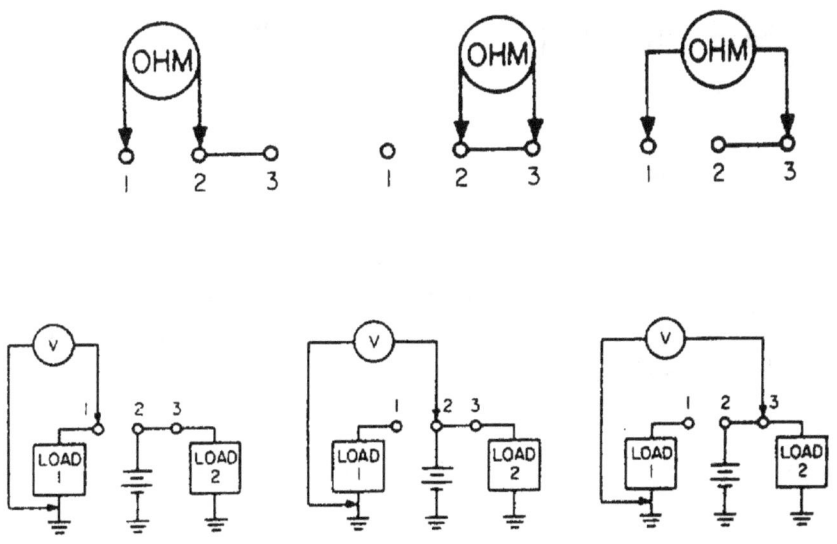

At least the same number of poles, throws, and positions; the same number of breaks and an identical configuration in regard to momentary and locked positions; and a voltage and current rating equal to or higher than the original switch. In addition, the substitute must be of a physical size compatible with the mounting, and must have the same type actuator as the original switch.

A **SOLENOID** is a control device that uses electromagnetism to convert electrical energy into a mechanical motion. The magnetic field of the coil and core will attract the plunger of a solenoid when current flows through the coil. When current is removed, the spring attached to the plunger will cause the plunger to return to its original position.

If a solenoid fails to operate, check the terminal connections, the plunger and attached mechanism for smooth operation, the energizing voltage, and the coil of the solenoid.

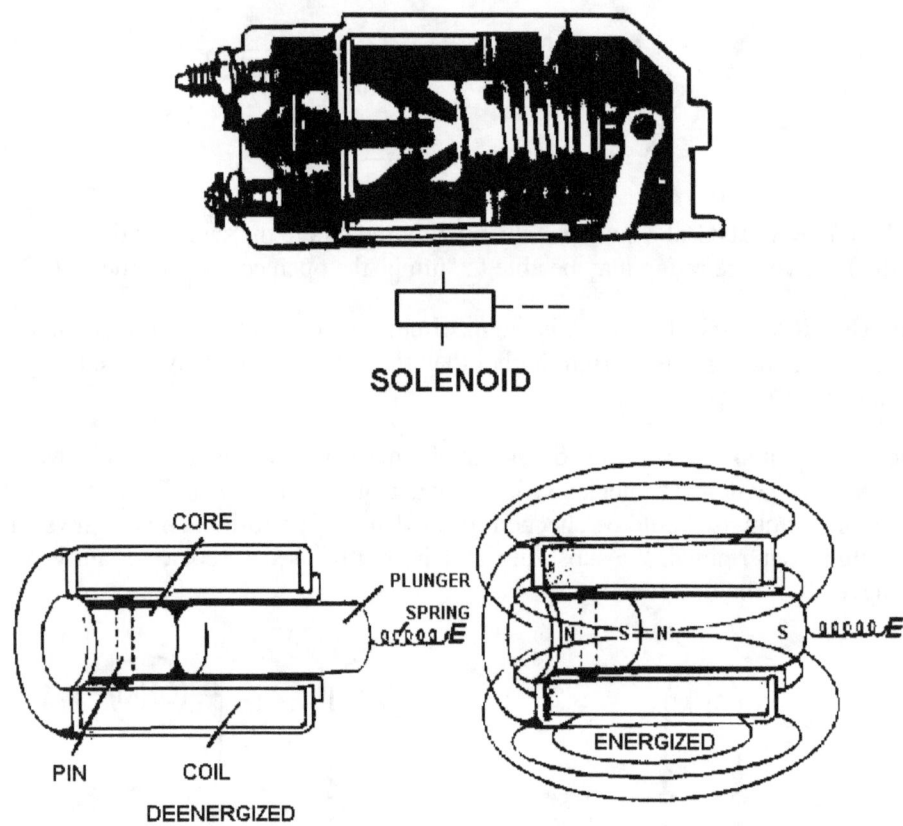

SOLENOID

A **RELAY** is an electromagnetic control device that differs from the solenoid in that the solenoid uses a movable core (plunger) while the relay has fixed core. Relays are classified as CONTROL RELAYS, which control low power COMMON circuits and POWER RELAYS or CONTACTORS which control high power circuits.

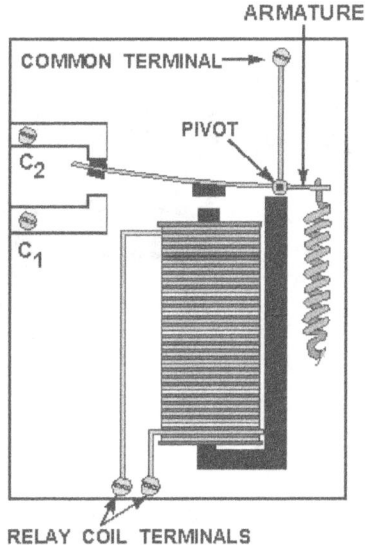

CLAPPER RELAYS use a clapper (armature) to move contact positions and accomplish the switching of circuits.

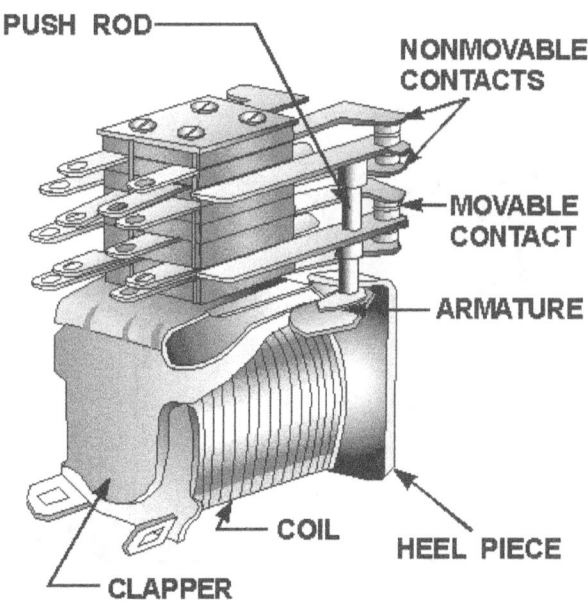

Relays are described by the type of enclosure. A relay may be OPEN, SEMISEALED, or SEALED.

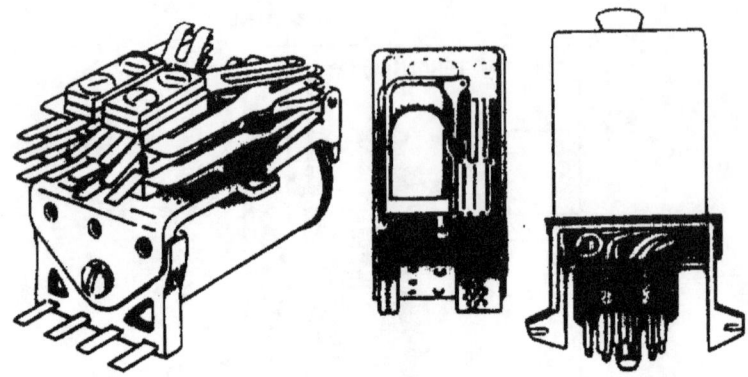

If a relay fails to function, the movement of the contacts should be observed; the coil should be checked for opens or shorts; the terminal leads should be checked for burned or charred insulation; and the contact surfaces should be checked for carbon, arcing, and contact spacing.

A **BURNISHING TOOL** is used to clean the contacts of a relay. Files, sandpaper, and emery cloth should NOT be used.

A **POINT BENDER** is used to adjust contact spacing of a relay. No other tool should be used.

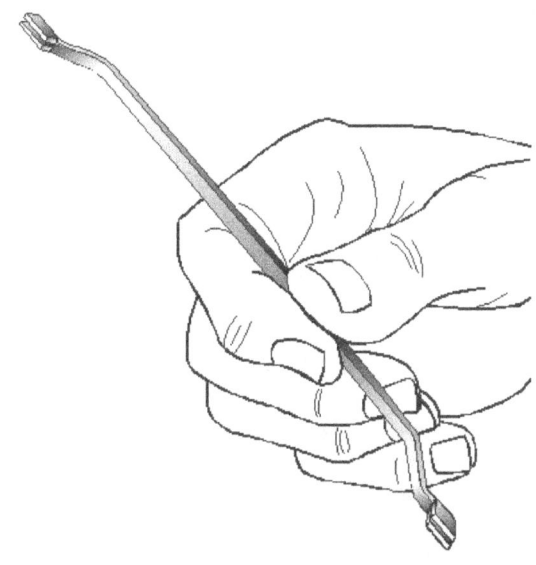

ANSWERS TO QUESTIONS Q1. THROUGH Q30.

A1. *To remove power from a malfunctioning device; to remove power from a device you wish to work on and restore power when the work is completed; to turn devices on and off as the device is needed; to select the function or circuit desired within a device.*

A2. *Switches, solenoids, and relays.*

A3.

    a. Solenoid.

    b. Switch.

    c. Relay.

A4. *A manual switch must be turned on or off by a person. An automatic switch turns a circuit on or off without the action of a person (by using mechanical or electrical devices).*

A5. *A light switch, an ignition switch, television channel selector, etc.*

A6. *A thermostat, an automobile distributor, a limit switch, etc.*

A7. *Multicontact switches make possible the control of more than one circuit or the selection of one of several possible circuits with a single switch.*

A8.

    a.   *Three-pole, single-throw (triple-pole, single-throw)*

    b.   *Double-pole, double-throw*

    c.   *Single-pole, double-throw*

    d.   *Single-pole, single-throw*

    e.   *Double-pole, triple-throw*

    f.   *Six-pole, double-throw*

A9.

    a.   *Single-pole, single-throw, single-break*

    b.   *Single-pole, double-throw, single-break*

    c.   *Single-pole, single-throw, double-break*

    d.   *Single-pole, double-throw, double-break*

    e.   *Rotary*

    f.   *Wafer*

    g.   *Double-pole, double-throw, double-break*

A10. *The type of actuator.*

A11. *Two-position and three-position.*

A12. *A momentary switch.*

A13. *A locked-position switch.*

A14. *A microswitch.*

A15. *The maximum current a switch is designed to carry.*

A16. *The maximum voltage allowable in the circuit in which the switch is installed.*

A17. *An ohmmeter and a voltmeter.*

A18. *A voltmeter.*

A19.

     a. *Not acceptable-single throw.*

     b. *Not acceptable-double break.*

     c. *Acceptable-choice #2 (different actuator).*

     d. *Not acceptable-single pole.*

     e. *Not acceptable-no momentary position.*

     f. *Acceptable-choice #1 (higher rating).*

     g. *Not acceptable-locked position incorrect.*

     h. *Not acceptable-current rating too low.*

     i. *Not acceptable-voltage rating too low.*

A20. *The switch operation for smooth and correct operation, the terminals for corrosion, and the physical condition of the switch.*

A21. *The magnetic field created in a coil of wire and core will attract a soft iron plunger when current flows through the coil.*

A22. *A starter motor and solenoid.*

A23. *The connections, the plunger, the mechanism that the solenoid actuates, the energizing voltage, and the coil of the solenoid.*

A24. *The magnetic field created in a coil of wire will attract aft armature causing a movement in sets of contacts.*

A25. *The solenoid provides a mechanical movement of a plunger (a moveable core) while the core of a relay is fixed.*

A26. *Control relays and power relays (contactors).*

A27. *By observing the movement of the contacts if the relay is open or sealed with a transparent cover. If the relay has an opaque cover, you can "feel" the operation of the relay by placing your finger on the cover.*

A28. *The coil should be checked for opens, shorts, or a short to ground; terminal leads should be checked for charred or burned insulation; the contact surfaces should be checked for film, carbon, arcing, and contact spacing.*

A29. *A burnishing tool.*

A30. *A point bender*